Yes!
Top Bar Hives

by

J. R. Slade

ISBN: 978-1-912271-22-1

Published in 2018 by:
Northern Bee Books
Scout Bottom Farm
Mytholmroyd
Hebden Bridge
HX7 5JS (UK)

www.northernbeebooks.co.uk

Tel: 01422 882751

Printed by Lightning Source, UK

Introduction

Top bar hives have had a bad press and many an established beekeeper has seized upon that bad press to damn them as not worth the effort.

In today's beekeeping world there is a drive to achieve a return for the beekeeper. Little or no thought is given to such things as the demise of feral bees, density of bee colonies in unitary areas, the nature of the food source etc.

Most beekeepers' thoughts are on getting more productive bees, varroa free bees, bees that don't sting etc.

We are progressively drawn to medication as a cure of all ills. We are told that there is a better bee strain and all we have to do is muck about with our queens.

We are continually having our attention drawn to "Emperors new clothes". Like the fairy tale they do not exist. Man, for all his skills, has never bettered nature when attempting to improve any live stock. Yes, we have cows that produce more milk than they did in the past but they are kept going like so many other domesticated animals with specialised housing, drugs etc.

How many of these animals would survive as they are, without the intensive input of modern farming?

We do not seem to understand that bees are not domesticated, they have to exist in the harsh reality of nature.

We take for granted that honey has wonderful healing properties, properties that bees primarily derive for themselves from plants that they have a symbiotic relationship with. Those same plants may not be appreciated by us, therefore we destroy them. Conversely other plants that are to our liking, showy, spectacular flowers from overseas that offer nothing to our bees what-so-ever, are abundant in our midst.

There are many beekeepers that say a better bee is breed-able.

What abject nonsense.

If we want to have bees that can exist and thrive in a changing world, the only hope is that we are sensible enough to stand back and let the bees do it for themselves.

We import bees from warm climates and cross them with other bees from unknown places and think that is good.

We delude ourselves.

We must take the opportunity that still exists, take a back seat and let bees swarm, produce drones without restriction and let them cross breed as they see fit. Let us put bees back in to the wild by making provision for feral colonies, which face the rigours of nature and offer a background of naturalised wild, native bees that will re-vitalise our stocks by reducing the contamination of imported strains.

Top Bar Hives offer a hope in beekeeping that does not exist in conventional beekeeping, where the emphasis is on the beekeeper's reward, rather than on the bees.

TBHs offer that freedom to bees (if un-interfered with) to create the comb they want, generate drones in the numbers that they see fit, to ensure that bee generation upon bee generation has the antecedence that they want and need to cope with the rigours of change.

Bees must be our priority not honey. The salvation of bees lies in regional mongrel-ism not in hybridisation.

The urban fox will always be there when the inbred pooch has gone.

We are frightened by the prospect of the Asian hornet getting here! Perhaps the problem with the Asian hornet, is that we didn't try to improve it. It copes in a way our bees do not.

We fail to understand that only nature mends, that only nature adjusts, that only nature gives answers, and that it will be nature that gives salvation where we cannot.

Top Bar Hive Beekeeping

Before setting out to produce top bar hives it might be best to recap on the fundamentals of conventional hive beekeeping.

With conventional hives everything that is relevant regarding the basic principles is built into the design of the hive and it's inner components. Everything from bee space to frame spacing is catered for in the manufacture of the component parts. Brood boxes are the right depth as are supers, queen excluders fit perfectly, so on and so forth.

When introducing a swarm into a conventional hive bees have no room to manoeuvre. They are constrained by the sheets of foundation. They have no other choice other that to use and draw out the foundation with whatever form is embossed upon it whether it be worker or drone.

Consequently the beekeeper has little to do other than observe and adhere to the guidance as set out in whatever bee book he or she has at hand.

I do appreciate that this is a simplification of the reality but none the less, little is left for the bees, or beekeeper to do other than stand back and watch the marvel of the bees work.

Top Bar Hive beekeeping comes with far less or no automatic bee constraints and the input of the beekeeper becomes paramount in ensuring that the bees have a home to their liking and the beekeeper has a hive that can be managed without the reliance on the resources of the hive manufacturers and their hive components, especially possibly detrimental materials such as recycled wax in the form of foundation.

I am sure that the purveyors of foundation do their best in cleaning recycled wax but we all know when purchasing foundation there is a great deal of variation in it's colour. That colour variation is a good indicator that what is being purchased is not the pure, slightly off white to pale yellow wax as it should be, but a substance containing all forms of residue and chemicals introduced to hives in previous usages.

Recycled wax has no traceable history, some foundation may come from a single recycle but most likely it is a mix of multi-recycles.

The basic fundamental of TBH beekeeping is that an individual hive contains nothing other than the wood from it's construction, an inert trigger membrane for comb building (starter strip**) and materials created by the bees themselves. (** starter strips will be covered further into the notes).

A Top Bar Hive to all intents and purposes is an empty box with four sides, a top and a bottom. The sides maybe set at an angle; in my design they are set at 15 degrees. The ends are usually vertical and constitute the construction anchors, ie the long sides are fixed to the ends. The bottom may or may not have removable slide out panels. By the very nature of the way in which bees behave in TBHs they appear to more readily control varroa and the necessity for any form of mesh screen in not there.

Some beekeepers may question this, as mesh screens have become part of what is now deemed as best practice. Varroa like nosema are considered to be things that have to be eradicated at all costs.

This is not an absolute, bees can and do survive with varroa and nosema might be considered a beekeepers own goal in that, quite often it is a feature of the way in which bees are managed.

Now we come to the top, the bars. In my book Why Not Top Bar Hives? I covered the difficulties in using full width bars that constitute a solid upper surface that the bees have to build comb down from.

View into empty TBH with end removed.

If there were no starter strips then the bees would select whatever start point they wanted and draw comb at whatever angle they wanted. The use of starter strips does help by giving the bees a launch pad so to speak.

One of the interesting things noticed when observing what the bees do in an empty, flat roofed space is that they appear to require a minimum "comb length" at the outset. This minimum comb length is to some extent a feature of the size, strength / vigour of the colony. It is this "bee requirement" for a minimum initial length of comb that must be considered when placing a swarm into an empty hive, and or expanding a colony.

If the space is too great, even with starter strips the bees will build comb in whatever direction suits them. Reducing the space such that they can only get the minimum "comb length" they want in the direction you want ie along the starter strip is imperative.

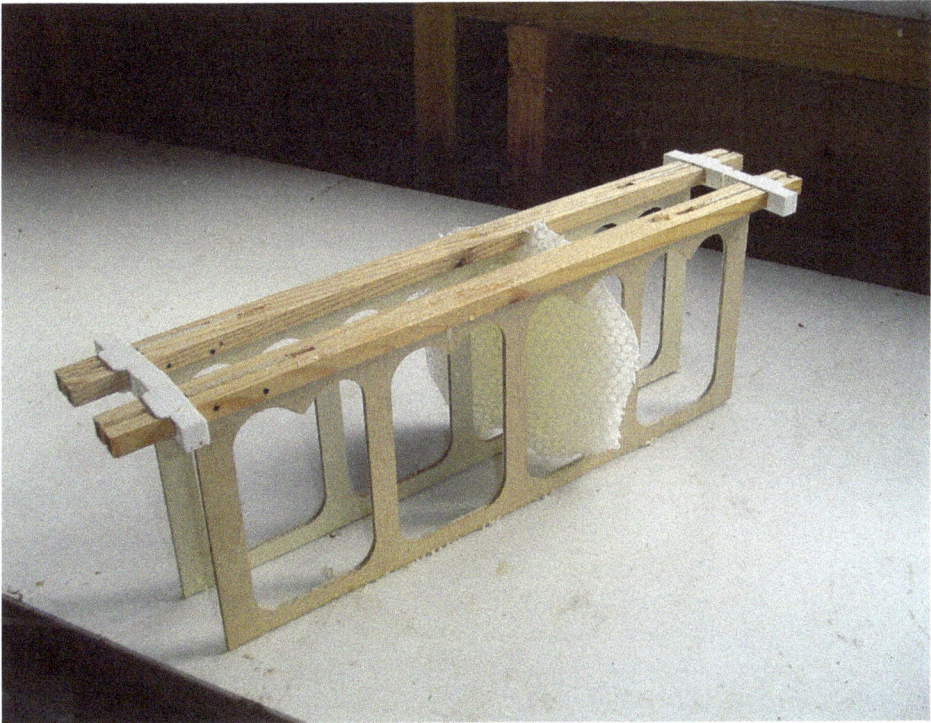

This picture shows what they do when given space in the wrong direction and have a restricted space in the right direction.

This picture shows that they will conform if only given space in the direction that you want. However when bees are forced to conform unnecessarily it is neither good for the bees or the beekeeper.

Bees that are doing their own thing (or think that they are doing their own thing) tend to be happier bees and are far less defensive.

It can be seen from the photograph of the completely empty TBH that the bees would have carte-blanche with how and where to commence comb construction.

Whereas in the photograph showing the divider board in place the bees are restricted, not only in volume available to them but the width of the space.

With four top bars in place, each having a width of 37mm (standard narrow spacer) the bees only have a 4 x 37mm = 148mm width to work in. From observation bees prefer to have a minimum "comb length" space greater than 150mm.

When placing a swarm into an empty hive, the number of top bars available to them is also dictated by the size of that swarm. A very large swarm may require 5 top bars, whereas a small swarm may only require 2 or 3 top bars.

The cross sectional area available to the bees of a TBH (my design) is approximately 1000cm2. Therefore the volume available below a single top bar is 3.7cm x 1000cm2= 3700cm3 = 3.7 litres.

Four top bars offer a volume of 14.8litres, five offers 18.5 litres and six offer 22.2litres and so on.

View into TBH with a divider board in place restricting the volume available to the bees and the "minimum comb length" ie the distance from the divider board and the end of the hive. A swarm will generally prefer to cluster in a single mass when placed in a new hive and provided they are not over crowded will remain as one. Observation in the days after placing the swarm in will tell you if you have it right. There is no harm in removing a top bar if you think that you have been too generous with space, or adding a bar if you find the bees spilling beyond the divider board.

These volumes may appear small however if the volume available at the outset is appropriate to the swarm size then the bees will readily draw comb along the starter strips.

Once it has been established that the bees are content in using and dropping comb from the starter strips the volume can be increased by the addition of further top bars.

Caution must be taken however when expanding with additional bars, as swarms tend to be eager beasts and if well fed (as a swarm should be) they may confound you by going beyond the divider board where they will produce comb to suit themselves, similarly if they have too much space before the divider board, again they will upset your plans.

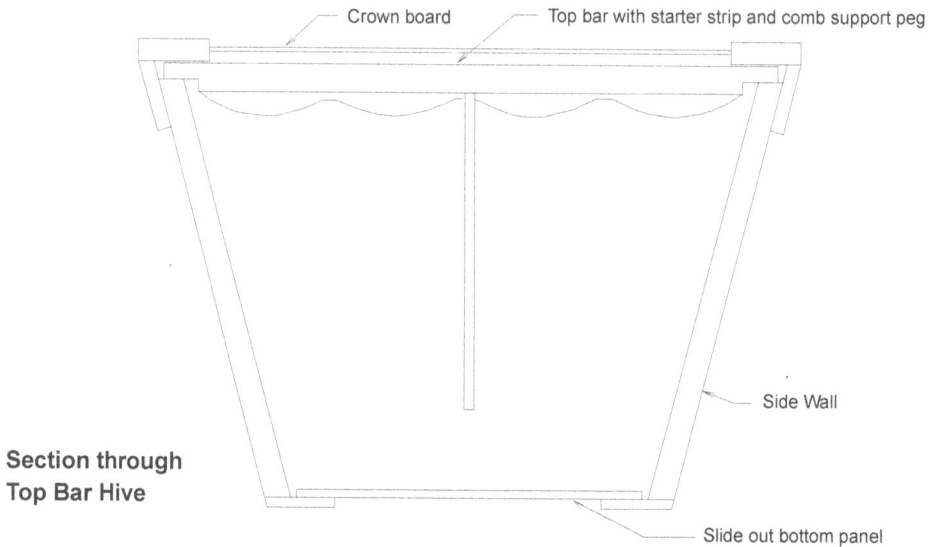

Crown board   Top bar with starter strip and comb support peg

Side Wall

**Section through
Top Bar Hive**

Slide out bottom panel

Starter strips

One of the most important questions in TBH beekeeping is "can we do away with foundation"?

The answer is a positive yes, bees are not only happier without foreign wax but start comb building at any start point and given any void they will commence building comb with gusto. It might be to one side or end of that void, it might be in the centre of that void. The one thing that is almost certain is that the initial section of comb will have a width exceeding 150mm. Having established that first piece of comb they will produce more comb that supports that piece whilst providing the "bee" and ventilation spacing they need.

The object of a starter strip is two fold; firstly getting the bees to place comb in a natural manner and secondly in a way that makes examination and manipulation of the colony possible.

If foundation or wax coated material is not to be used as a starter strip, what can be used?

The most important requirement is that what ever is used, it is not detrimental to the bees. Additionally the material must be suitable for contact with food for human consumption. The material should also not have any character that the bees might not like, such as having a chilling effect as metal might have.

This reality leads to one or other of two materials; wood or plastic. Wood is suitable in many ways but it does need to be thin ie less than 3mm thick and stable. This presupposes plywood. There is a down side to plywood in that it does contains an unknown component in the form of the adhesive, bonding the plies together.

From observations the bees will use a plywood starter strip to draw their comb down, however they prefer not to draw comb out from the flat surface of the plywood other that to give a better anchor at the point of contact on the starter strips lower edge.

This can be clearly seen on the photograph below.

Note the comb has a minimal adherence area onto the plywood frame, to the point that where it is not essential there is no attachment at all.

If not wood then plastic, but what plastic? Most likely plastics fall into three categories, the first being a plastic that bees may not like.

Some plastics are quite dense and as such conduct heat, not necessarily as readily as metal but too much.

Secondly plastics that might be detrimental to bees. This is a greyer area and whilst most plastics are fairly inert they may or may not or not be harmful and it is the unknown that must determine whether to use or not use. In other words play safe.

The third category are those plastics that are approved for use with food for human consumption, and do not fall into the first two categories.

There is a plastic in sheet form that does fall into category three. It's a foamed plastic of low density and therefore has a good thermal characteristic and has a slightly textured surface that bees appear to be happy with.

The problem with the foam is that it only comes in a 1220mm x 2440mm sheet and has be cut to the required shapes by sophisticated machinery.

Although the cost of the material is not great, neither is the cost of machining, when combined the cost becomes unreasonable for a beekeeper with just a few TBHs.

This unfortunately brings us back to foundation, however it is available to any beekeeper and the bees will work with it.

Referring back to Why not Top Bar Hives?, the use of top bars that are narrow ie 22mm (not 37mm wide that abut) and the use of a central comb support peg, there are a couple of things that can further help in providing the bees with a better start.

Firstly the depth of the starter strip needs to be between 40mm and 50mm deep with a wavy lower edge, see diagram following.

The diagram following shows a sheet of national super unwired foundation cut into 8 pieces. The first cut is to halve the sheet into 2 strips left to right. The the 2 halves are halved again with a wavy cut left to right. Finally the 4 strips are cut in the centre top to bottom.

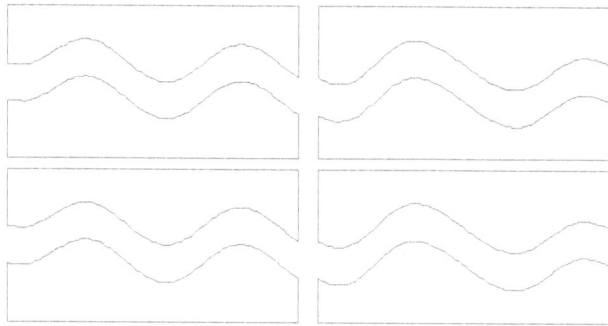

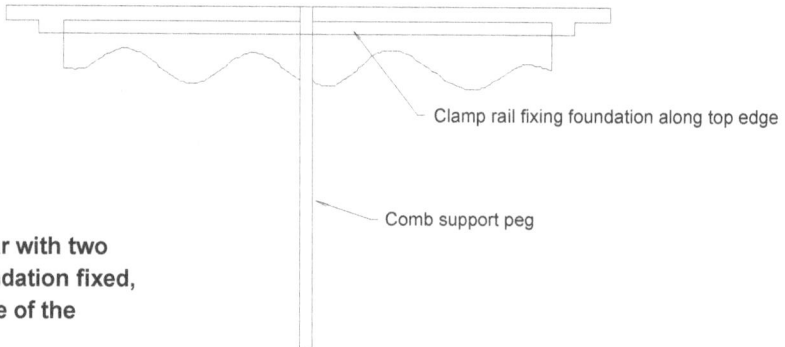

Clamp rail fixing foundation along top edge

Comb support peg

View of top bar with two pieces of foundation fixed, one either side of the support peg.

The following view shows the fixing of the support peg and foundation clamp rails, (if using a standard top frame rail it will be necessary to cut the clamp rail in two.

Note that if using a square peg it should be oriented so that two corners are in line with the foundation.

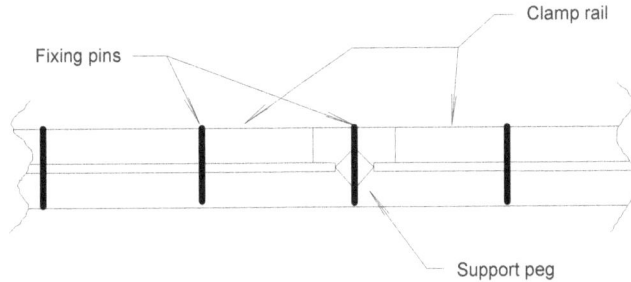

Clamp rail

Fixing pins

Support peg

If using standard frame top rails for the top bars, a cut down bottom frame rail can be used as the support peg.

Although the bottom rail is rectangular, it can be used by drilling a 6.5mm diameter round hole in the centre of the rail, whittling one end approximately round and driving it in and finally fixing with a gimp pin. It is very important that the peg is absolutely vertical in use. Bees will always draw comb down vertically and if the peg is not exactly in the centre of the comb it will not give the support that it should.

A simple way to make sure that the peg is as it should be, is to place the finished top bar with peg on its side (the side opposite to the foundation fixing rail down) on a flat surface and check that the peg runs parallel to the surface. See below.

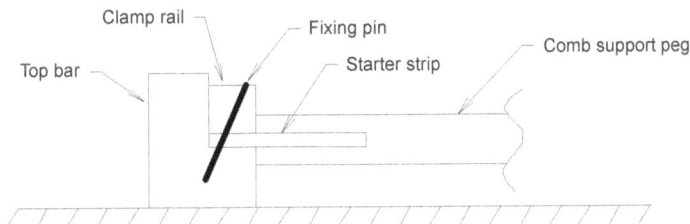

Clamp rail
Fixing pin
Comb support peg
Top bar
Starter strip

From speaking to other beekeepers that have tried TBHs and have found them a little testing, two things always come into the conversation. One, the fact that they have been misled by those that wish to sell them an inferior, un-thought out design, where the top bars are full width (37mm) and abut, thus making it almost impossible to clear bees from to top of them when closing the hive down. The other is the lack of comb support. With no support lifting and viewing a drop of comb suspended under the bar is not just impossible but dangerous especially to the bees. Should the comb separate from the bar it will crash to the floor or wherever with no hope of recovery.

The use of narrow top bars (22mm) with standard narrow spacers together with a crown board has been covered in Why Not Top Bar Hives? so there is no need to go over this point.

Getting bees into TBHs is something that has to be looked at.

There are a number of ways in getting bees into a new hive. One is to simply divide a strong hive, placing half of the comb into a new hive, ensuring that the new hive has a good number of nurse bees that will not leave the hive when moved to a new location. The queen does not have to be seen for this process providing there are fresh eggs (less than 2 days old) in both the original hive and the new hive. The hive that finds itself queenless will get straight to work producing queen cells and thus a new queen from the new eggs.

This process in the case of TBHs requires a TB donor hive.

From scratch the best and easiest way to get bees into a new TBH (especially if this is your first venture into TBHs) is by means of a swarm.

There are two fundamental ways in which to get a swarm into a new hive. One is the "American dump system". Take the lid off the hive, dump the swarm in and put the lid back on quick. This does work, but quite often, unless there is an obstruction in the form of a piece of queen excluder over the entrance, the bees will simply flood out of the front almost as quickly as you dumped them in the top.

For me the traditional method remains the best, letting the bees walk in to their new home of their accord.

Yes this can also go wrong, some times the bees will not go in. That being the case then you are the problem, as they do not like what you have provided. If the new hive is as it should be then the bees will always avail themselves of a good new home.

As said bees will always go into a good home but TBHs have a very small entrance compared with the entrance of say a national hive, that has an entrance 20 odd mm high and the full width of the hive.

With a TBH it is necessary to carefully set the ramp (I use a board with a white sheet covering) that the bees are going to walk up. If possible the white cloth should be taut and placed so that the highest point on the cloth is at the entrance.

Additionally to ensuring the ramp is properly in place when walking the bees in, it is necessary to remove the hive lid.

This is not because of some deep technical reason but because the sides of the hive are not vertical. This combined with the fact that the hive lid is of generous proportion, offers a significant shaded area above the entrance that the bees can mistake for their final destination.

Note that the white cloth carries the bees ever upwards toward the entrance.

What you can be sure of is that when bees enter a new hive of their own accord they are very unlikely to abscond.

A swarm as we know, will deprive the hive from which it emerged of a good amount of stores, none-the-less a swarm needs to create a large area of comb. Food placed on a swarm is greatly needed. To produce a pound of wax requires many pounds of food. Initially that food is not destined as larva food but simply to create wax.

A brief run through of the salient points of bees in TBHs

1. Abutting top bars are a no no.
2. The use of a comb support peg is an absolute imperative.
3. The use of narrow top bars with end spaces affording bee movement is essential.
4. The use of a crown board with bee space on the underside essential.
5. The use of an eke. There must be a ventilated space above the crown board. Bees produce heat and moisture, without that ventilated space the underside of the roof sweats and creates dampness in the top bars.
6. The entrance must be of modest size (25mm x 100mm)
7. The entrance must not be at the bottom of the hive neither must it be at the

end. Having the entrance adjacent to comb together with its modest size helps to prevent wasp invasion.

8. Appropriate starter strips in terms of depth, form and material are essential.

9. The long sides of the hive must not be vertical. From experience 15 degrees is about right.

10. When introducing a swarm, that swarm must have just about the correct volume to ensure that the bees work along the starter strips.

11. Regular observation is necessary (without interference) when expanding with additional top bars.

12. Do not ignore your new swarms. An eye as to how they are doing is essential in the first few days.

13. Remember you are working for the bees not they for you.

14. The most important thing of all to bear in mind is that bees are subject to Nature and Natures' apparent indifferent cruelty. Not all bees do well, not all bees survive but the fittest will and that is our hope. Bees buoyed up with chemicals and medicines etc, will in the long run, be our and their undoing.

A thank you to all those beekeepers that have had a go with top bar hives.
It is still early days and many things have to be resolved with the use of TBHs, as yet even the simplest things are not determined in respect of what is truly the best for our bees.

Hopefully there are those amongst us that will persevere and establish a background of TBH beekeepers, strong enough to determine constructional parameters and methods of management. This will be an imperative, so that any one wishing to keep bees in this way knows that what they are doing is for the betterment of bees and is not continuing the myopic view of bees being servants that have to conform to our will.

The cavalier methods of livestock keeping, where the nature of that stock is disregarded to service the convenience of the farmer is coming home to roost.

Will any of us be safe in ten years time from the effects of uncontrolled use of antibiotics in the farm yard? A simple infection might mean an individual's demise because antibiotics are progressively becoming ineffective from over-use, a use that is still deemed "good farming practice"

TBHs and their use will if nothing else give us the opportunity to question, some of the so called "good practices" proffered by the beekeeping establishment.

Notes and drawings

for the construction of a Top Bar Hive

To date, no one has come up with a definitive design for top bar hives. It is much to the discredit of hive manufacturers that they have not and only offer to the more thoughtful beekeepers what can only be described as products where no thought what-so-ever has gone into any form of standardisation.

Whilst not suggesting that what follows is a final answer it may be a constructive start in that if sufficient beekeepers can agree on some basic sizes etc then a standard will follow by default.

Introduction

There are four fundamental dimensions that have to be considered when constructing a TBH.

First:- <u>The length of the top bars themselves</u>. Within reason they can be any length provided that they are not too long. For convenience and also because it offers a length that is suitable and manageable I personally settled on a National frame top rail length which is 17 inches (432mm). Others may prefer a longer top bar and choose say a Langstroth frame top rail. By using a standard top rail there is the benefit of the "clamp section" of the top rail to fix whatever starter strip is chosen.

Another reason for using a national frame top rail for the top bars is that plastic spacers are readily available.

Second:- <u>The angle of the sides of the hive</u>. If the sides of the hive are vertical there is a tendency for the bees to fix any comb they create to the sides to the full depth of the comb. If the sides are angled too greatly, the volume is reduced too much and therefore the area of any given drop of comb is also greatly reduced.

An angle of 15 degrees seems to be about right. The bees do attach the drops of comb to the sides but only to a point about 2 inches (50mm) below the top bar. When manipulating this small degree of attachment is easily cut through to free up the comb for removal and examination.

Third:- <u>The depth of the hive</u>. There has to be a balance between the depth and the width at the bottom. When the hive is on a stand, is it wide enough to be stable? (You can see from photograph on front cover how a TBH sits on it's stand)

Presuming the use of 17 inch (432mm) national frame top rails, a side angle of 15 degrees and an internal hive depth of 12 inches (306mm) the horizontal dimension of the base is approximately 11 inches (280mm), wide enough to be stable on an appropriate stand.

I mentioned the use of a national frame top rail because it is the shortest length that I consider to be suitable for a TBH. Irrespective of top bar length, a depth of 12" remains sensible and with a longer top bar the width of the base is greater, therefore more stable.

Using Langstroth frame top rails and retaining an internal depth of 12" the base width goes up from approximately 11" (280mm) to 13" (330mm)

The forth:- The length of the hive. This dimension is the internal length plus the thickness of the end walls of the hive. The internal length is determined by how many top bars are used. The accepted width of a section of comb (with worker brood) is 1.46 inches (37mm) therefore if the hive is to have provision for 20 top bars, the internal length will be 20 x 1.46 = 29.2 inches. If this is increased to 30 inches, there is room for a false frame (dummy board) that can be removed to make manipulation easier. The overall length of the hive is therefore 30" (762mm) plus the wall thickness's. With 7/8" (22mm) end walls the length would be 31 3/4" (806mm) This could for convenience be rounded up again to 32" (813mm)

From experience a TBH that has 20 top bars is about right.

The length of the hive (nominal number of top bars) together with the length of the top bars is something that has to be resolved, if there is to be a standard TBH. We are in the situation at the moment where TBHs are being manufactured that do not conform to any pattern and no two manufacturers work to the same dimensions. If TBH use is to become an integral part of beekeeping, then some standardisation must exist. I hope that the suggestions in this leaflet will constitute a reasonable starting point.

For the construction of the body of the hive Western Red Cedar is recommended. Other internal parts such as crown boards can be manufactured from good quality external plywood

Diagram 1

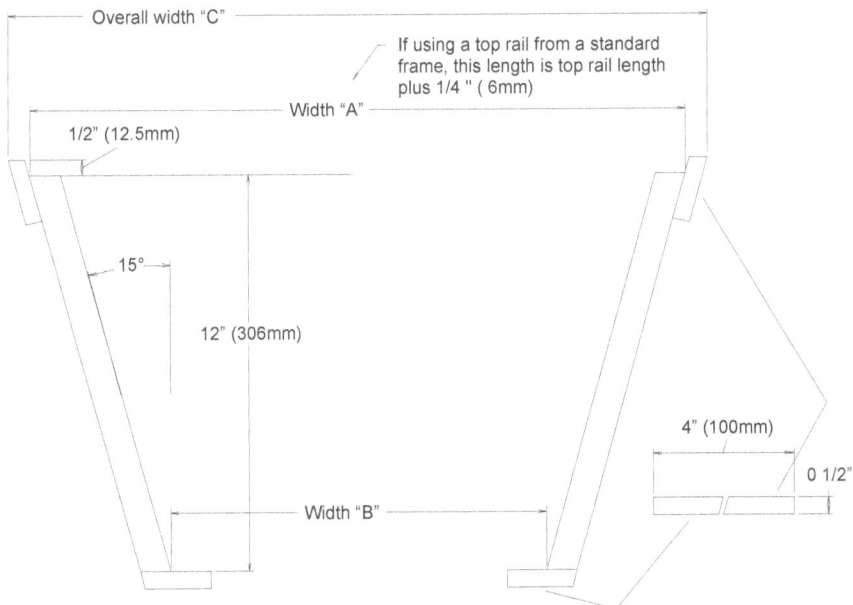

This drawing shows a cross section of a TBH with the fundamental dimensions that are necessary for construction. The width "A" will automatically give dimension B if the walls are set in at 15 degrees.

Note - all the drawings in this leaflet are semi-accurate illustrations only and should under no circumstances be scaled.

There is no dimension for the thickness of the walls. The thickness of wood normally used to construct a conventional hive is 7/8" PB(22mm) This thickness comes as a practicality; 1" planed.

The 1/2" (12.5) where the top bars rest will suit both National and Langstroth frame top rails. Although National rails have 3/8" (10mm) ends and Langstroth have 1/2" (12.5mm) ends, the difference in use will be of no concern. It will increase the bee space under the crown board but not sufficiently to be a problem.

The 4 rails are manufactured as per the inset and are pinned and glued in place. The lower rails constitute the support for the removable panel.

Diagram 2

Alighting board

The drawings above show an exploded open end view with a top bar and the alighting board in place.

The drawing below shows the construction of the alighting board.

Diagram 3

3" (75mm)

Fixing hole screw

3" (75mm)

1/4" (6mm)

3/4" (19mm)

5" (125mm)

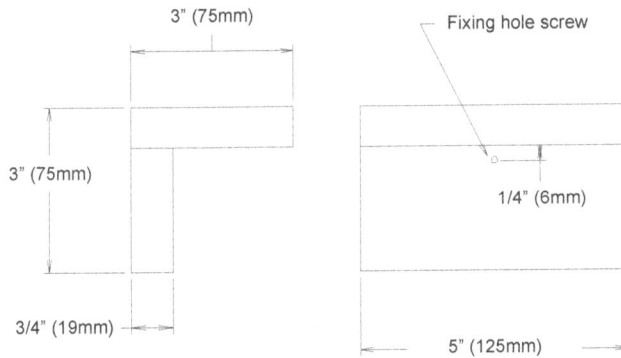

Note the position of the fixing screw. It is placed such that if required it can be slackened slightly and the whole alighting board rotated 180 degrees to act as a closure for the hive if necessary.

The alighting board needs to made from Western Red Cedar.

Below is an exploded / truncated side elevation showing the fixing of the end walls, the removable panels and position / size of the entrance slot.

Diagram 4

Length "L" See introduction

7/8" (22mm)

6 x No 8 x 2 1/2 Brass screws each end

Entrance slot

End wall

4" (100mm)   3 1/2" (90mm)

There are two slide-in removeable floor panels that just touch when fully pushed in

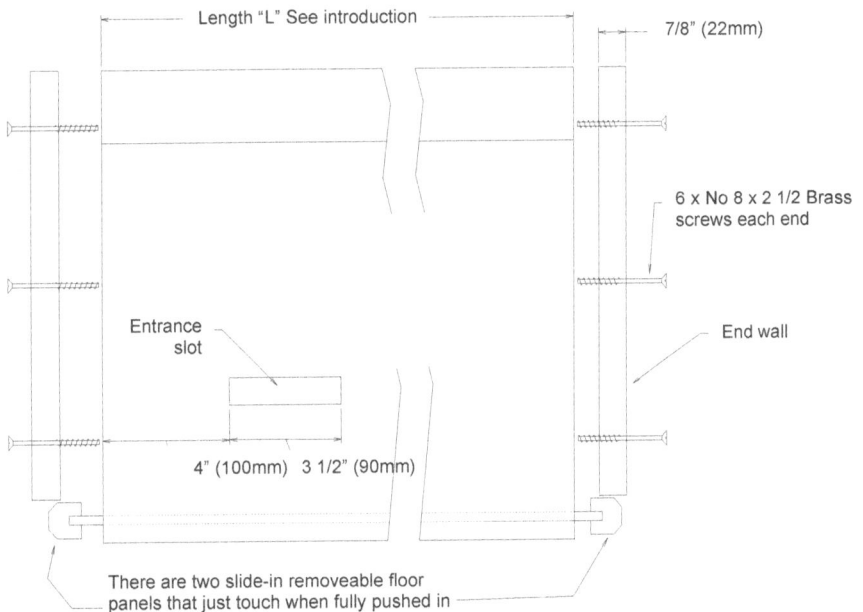

The entrance slot may seem a little small to some beekeepers, however the reality is that the smaller the entrance that the bees can manage with, the better. Too small and bees have leaving and returning problems, too large and the bees have defence and intruder problems, especially in the wasp season.

Below is an end elevation of a hive, together with a side elevation of the end wall showing the clearance under the end wall for the removable panel.

Diagram 5

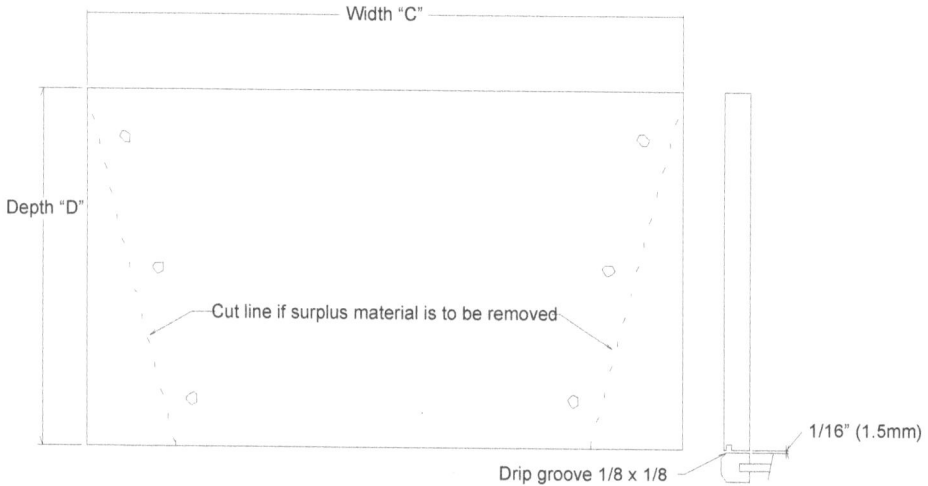

The drawing also shows the approximate positioning of the end wall fixing screws. It is not necessary to remove surplus wood and in some instances a broader base may be an advantage on some stands.

The photograph on the front cover shows a hive resting on cross rails. Had the surplus not been removed the hive end walls could have rested on the longitudinal rails. When making a determination whether to remove or leave the surplus on, provision has to be made to slide out the removable panel at both ends.

Diagram 6

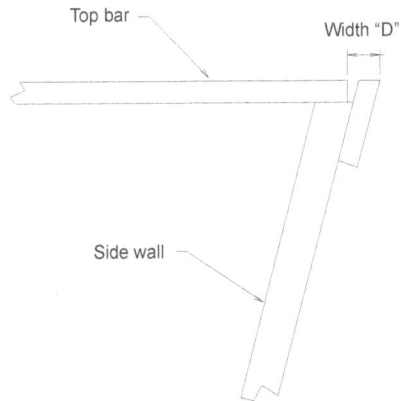

The Width "C" ie the overall width of the hive is the length of the top bar, plus 2 x Width "D"

Width "D" is calculated thus.

1/2" (12.5mm) divided by cos 15 degrees, plus 1/2" (12.5mm) x tan 15 degrees divided by 2, plus top bar end clearance of 1/8" (3mm)

Resolved this is 1/2" (12.5mm) divided by 0.966, plus 1/2" (12.5mm) x 0.0.134, plus 1/8" (3mm)

Which gives 17/32" (13mm), plus 1/16 (2.5mm) plus 1/8" (3mm)
all equalling 23/32" (18.25mm)

To make things easier, assuming the hive side rail is 1/2" (12.5mm) plus a bit and the depth from the bottom of the top bar to the top of the side rail is also 1/2" (12.5mm) plus a bit, every thing can be rounded up to 3/4" (19mm),
The overall width of the hive Width "C" is top bar length plus 1 1/2" (38mm)

Depth "D" is simpler; it is, internal depth ie 12" (306mm), plus top bar thickness of 1/2", less, 1/16" (1.5mm), less "1/2" (306mm)
This is 12" + 1/2" - 1/16" - 1/2" - 1/8" (if panel is 1/4" thick)
= 11 13/16" (300mm)

Diagram 7

1/2"

Depth "D"

12"

1/16"

1/2"

1/8"

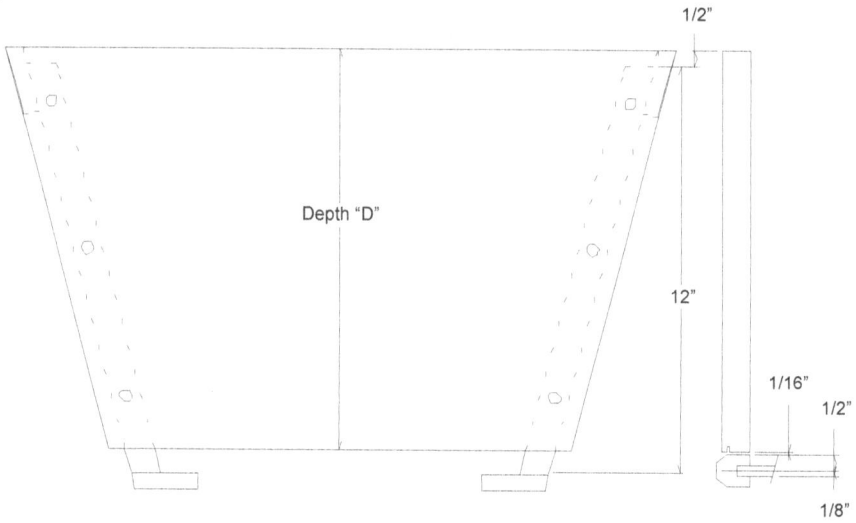

This diagram shows an end elevation with the removable panel missing. The detail to the right shows the dimensions used to calculate "D"

The removable panels comprise two rectangles of plywood and two lengths of 1" (25.5mm) square Western red cedar.

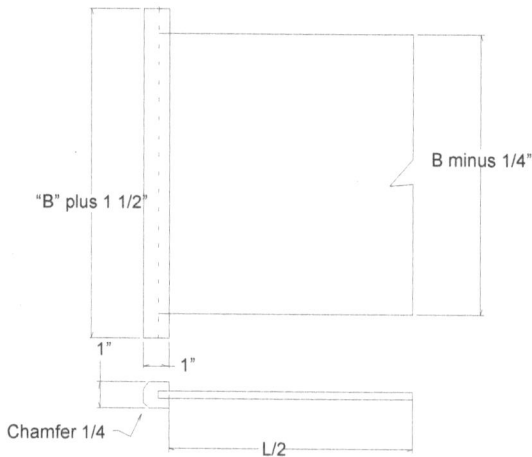

B minus 1/4"

"B" plus 1 1/2"

1"

1"

Chamfer 1/4

L/2

The Width "B" can be calculated from the hive overall width and the internal depth.

From the following diagram it can be seen that the width "B" is overall hive width "A", minus 2 x ("Y" plus "X")

"Y" is 12 1/2" (317 5mm) x tan 15 degrees which is 12 1/2" (317,5mm)x 0.27 = 3 3/8" (86m)

"X" is the thickness of the hive side rail and hive wall thickness, ie 1/2" + 7/8" (12.5 +22mm) plus a bit (say 1/16" 1.5mm) = 1 7/16" (36mm)

Therefore "B" is Width "A" minus 2 x (3 3/8" + 1 7/16") = 2 x 4 13/16"(122mm) = 9 5/8" (243mm)

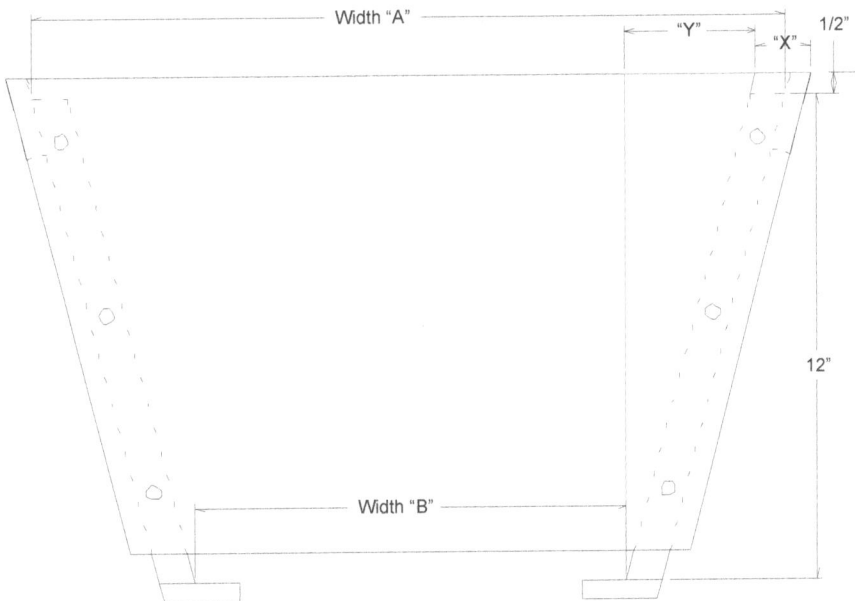

Having settled on the size of the hive body the next items are the crown boards.
In "Why Not Top Bar Hives?" I have shown three crown boards of different lengths. This may not suit every TBH beekeeper. Some may want a single large one others may wish to stick with me and have three. Here I will assume double crown boards both to the same dimensions.

Following is a diagram showing the construction of a crown board.

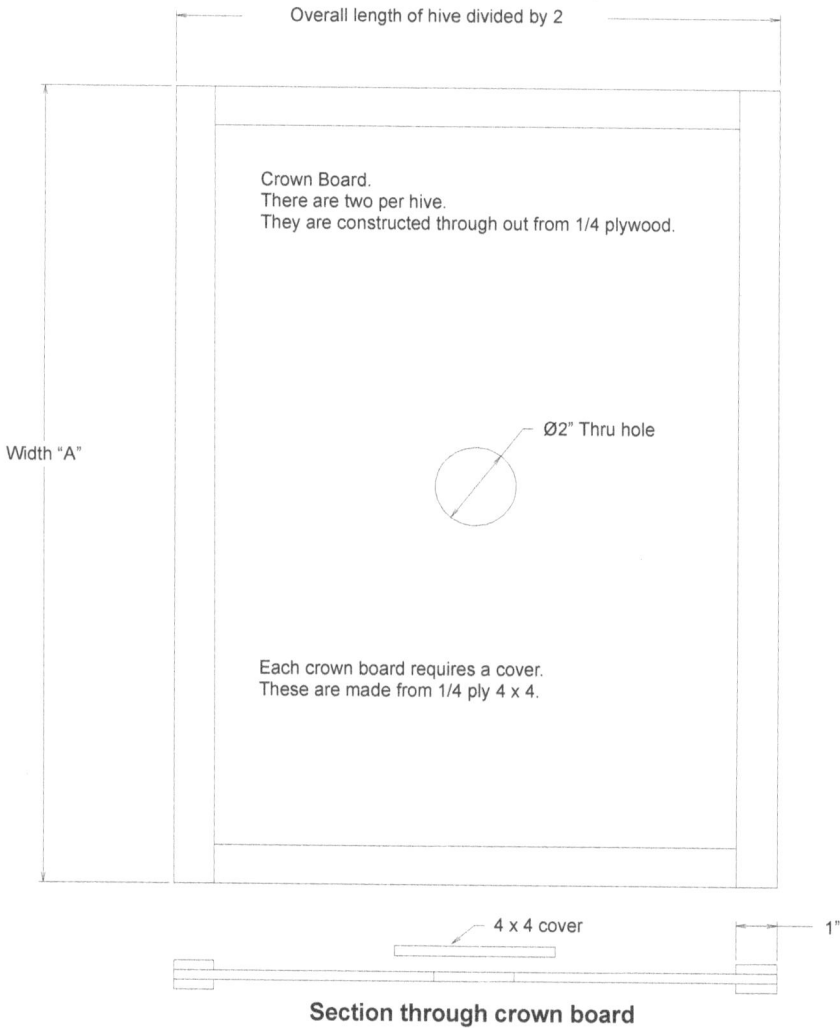

Overall length of hive divided by 2

Crown Board.
There are two per hive.
They are constructed through out from 1/4 plywood.

Ø2" Thru hole

Width "A"

Each crown board requires a cover.
These are made from 1/4 ply 4 x 4.

4 x 4 cover

1"

Section through crown board

The hole in the centre of the crown board permits the use of a contact feeder. Unless the ambient temperature is very high the hole is normally closed off with the 4" x 4" cover.

The following diagram shows the construction of an eke.

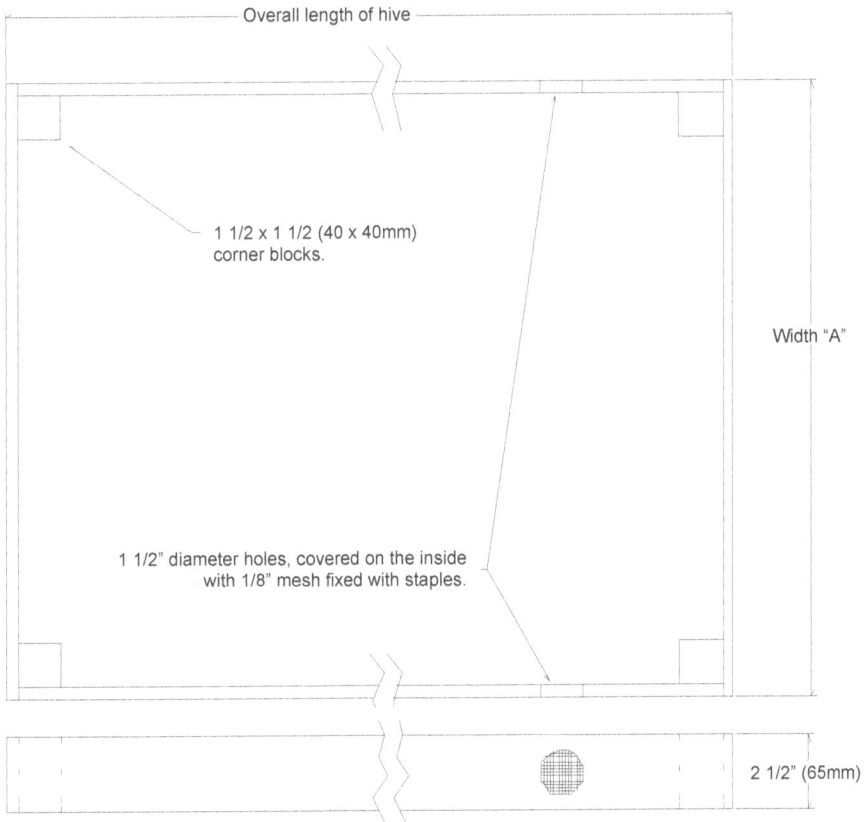

Overall length of hive

1 1/2 x 1 1/2 (40 x 40mm) corner blocks.

Width "A"

1 1/2" diameter holes, covered on the inside with 1/8" mesh fixed with staples.

2 1/2" (65mm)

The material for the eke can be either solid ½ x 2 ½" board or 3/8" (10mm) plywood.

The use of an eke is essential to ensure that the space above the crown boards is ventilated. The ventilation is needed in the summer to prevent the hive over heating, when the cover on the crown board can be removed. In the damp part of the season ventilation will prevent moisture build up, that might contaminate the crown board and top bars. The mesh covered holes may be set in the long sides or the ends.

The roof construction does not need to be set down but there are a few things that must be kept in mind.

The internal dimensions must be generous enough so that removing and replacing can be done easily, ie approximately 2" (50mm) greater than the maximum dimensions of the hive.

The vertical section / walls must be deep enough to extend down well below the crown board, 5" (125mm) minimum.

The material for the construction must be durable; even with a waterproof covering it is the component most exposed to the elements. Sterling board measuring 3/8" (10mm) is very good but its lower cut edges must be sealed with a varnish.

The water proof cover must be one that does not add too much weight. Mineral coated tar type roofing felt is fine.

Finally the roof does need handles. See photograph on front cover.

There are two other simple things that have to be taken into account when looking at TBH design, namely, feeders and dummy boards.

The use of a contact feeder is possible if there is space created by the use of an eke. Assuming that there is an eke as proposed then a surface contact feeder can be used. However, the common "rapid feeder is too deep to fit within the eke space but there are slimmer feeders called "nucleus feeders"; these are approximately 2" (50mm) deep. These are similar to the rapid feeder but hold a little less syrup.

Whatever type of feeder is used with a TBH it must be understood that the use of large volume feeders is not possible, unlike the case with conventional hives.

I personally question the use of these feeders; are they for bee feeding or are they to provide bees with an abundance of syrup that they can convert to a type of honey to be passed off for something that it is not? For the non-commercial beekeeper I cannot think of an instance when 30 plus pounds of syrup needs to be given at one time to a single colony of bees.

The other option for small quantity bee feeding is a "drop in" feeder.

The following diagram shows the construction of a "drop in" feeder.

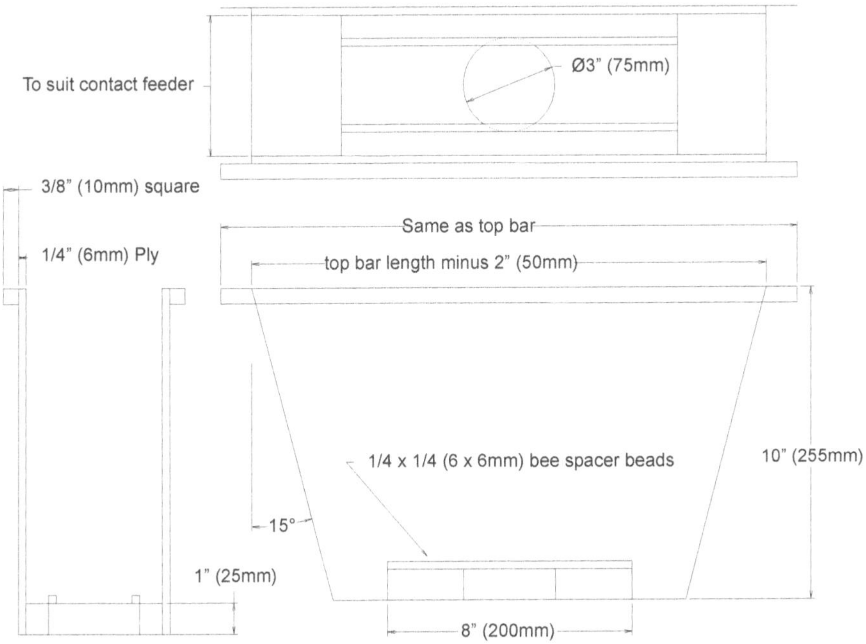

To suit contact feeder

Ø3" (75mm)

3/8" (10mm) square

1/4" (6mm) Ply

Same as top bar

top bar length minus 2" (50mm)

1/4 x 1/4 (6 x 6mm) bee spacer beads

10" (255mm)

15°

1" (25mm)

8" (200mm)

The construction is basically two sheets of 1/4" (6mm) plywood shaped as shown, suspended by means of two 3/8" (10mm) wooden rods. The two sheets are then attached to each other by means of a block of wood 1" (25mm) thick that has a central hole 3" (75mm) diameter in the centre. Additionally there are two 1/4" square beads that the contact feeder rests on. Details for the "drop-in contact feeder" itself can be found in "Why Not Top Bar Hives".

The other little thing is a dummy board.

The construction of a dummy board is the same as the construction for one of the sides of the drop in feeder.

There is no standard for a TBH as yet but there has to be a start somewhere. All of the formulations so far in this book have been based upon a top bar of an undetermined length.

With my TBHs I have taken a National frame top rail as my standard. If there is to be a standard then the best option will be, to use either a National or Langstroth frame top rail as the standard that fixes all of the principle dimensions of the hive. Then comes the length. The probability is that TBHs will be the domain of two primary users; amateur beekeepers and beekeepers whose main interest is in providing a service as pollinators, ie those that are not intent necessarily on a honey crop but want ease of construction of their hives and convenience in use.

For the amateur, a 20 top bar hive is a sensible and realistic size, not too big or too small. Whereas for a pollinator beekeeper a much smaller hive, ie a 10 top bar hive might be be considered.

Some constructional and dimension details might need adjusting but it would give a robust hive in terms of bee mass.

The construction and dimensional details for a "pollinator hive" will have to come at another time. For now the best thing is to make things as easy as possible for the amateur.

Using a National frame top rail (17" 432mm) long, other derived dimensions are:-
Overall width- 18 9/16 (472mm)
End wall would therefore be, 18 9/16" wide x 11 13/16" deep (472 x 300mm)
Width of removable panel 8 3/4" (222mm)
Length of rod on removable panel 10 5/8" (270mm)

With 20 top bars:-
Overall length 32" (812mm)

Crown boards, assuming 2:-
Width 18 9/16" (472mm) x 16" (406mm)

Roof:-
Internally, width 20 9/16" (522mm) x length 34" (866mm) x depth 5" (125mm)

For a Langstroth frame top rail, all width dimensions are increased by 2" (55mm) Length dimensions together with the end wall depth remain the same.

The Photograph at the start of this section shows a TBH with multiple entrances. It was designed to obtain information about entrance size and position. The entrance on the hive on the front cover is 1/2" x 2 1/2"(12.5 x 65mm) which in fact proved to be a little on the small size, similarly the alighting board was also a little too small.